TABLE OF CONTENTS

		page
I.	Introduction	7
II.	Issues	13
III.	Conclusions and Recommendations	15
IV.	Combustion in Coal Fired Units	19
V.	Results of the Field Investigation	27
VI.	Present State of Research	31
VII.	Economics	41

Appendices

A.	Principal Interviews Conducted	51
B.	Selected Bibliography	54
C.	Glossary	61

List of Exhibits

I.   Imports versus Use   8

II.   OECD Potential Coal Use by Sector   9

III.   OECD Projected Steam Coal Imports   10

IV.   Hard Coal Production by Region/Countries   11

V.   Fuel Prices for Electricity Generation   12

VI.   Coal Characteristics vs. Effects   16

VII.   Elevation of a Typical Coal-fired Boiler   22

VIII.   Effect of Heating on Minerals in Coal   24

IX.   Forced Outage Rate versus Coal Quality, Japan   45

me

INTERNATIONAL ENERGY AGENCY

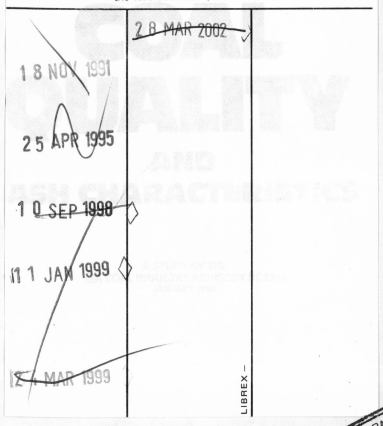

INTERNATIONAL ENERGY AGENCY

2, RUE ANDRÉ-PASCAL 75775 PARIS CEDEX 16, FRANCE

The International Energy Agency (IEA) is an autonomous body which was established in November 1974 within the framework of the Organisation for Economic Co-operation and Development (OECD) to implement an International Energy Program.

It carries out a comprehensive programme of energy co-operation among twenty-one* of the OECD's twenty-four Member countries. The basic aims of IEA are:

  i)   co-operation among IEA Participating Countries to reduce excessive dependence on oil through energy conservation, development of alternative energy sources and energy research and development;

  ii) an information system on the international oil market as well as consultation with oil companies;

  iii) co-operation with oil producing and other oil consuming countries with a view to developing a stable international energy trade as well as the rational management and use of world energy resources in the interest of all countries;

  iv) a plan to prepare Participating Countries against the risk of a major disruption of oil supplies and to share available oil in the event of an emergency.

*IEA Member countries: Australia, Austria, Belgium, Canada, Denmark, Germany, Greece, Ireland, Italy, Japan, Luxembourg, Netherlands, New Zealand, Norway, Portugal, Spain, Sweden, Switzerland, Turkey, United Kingdom, United States.

Pursuant to article 1 of the Convention signed in Paris on 14th December, 1960, and which came into force on 30th September, 1961, the Organisation for Economic Co-operation and Development (OECD) shall promote policies designed:

  – to achieve the highest sustainable economic growth and employment and a rising standard of living in Member countries, while maintaining financial stability, and thus to contribute to the development of the world economy;

  – to contribute to sound economic expansion in Member as well as non-member countries in the process of economic development; and

  – to contribute to the expansion of world trade on a multilateral, non-discriminatory basis in accordance with international obligations.

The Signatories of the Convention on the OECD are Austria, Belgium, Canada, Denmark, France, the Federal Republic of Germany, Greece, Iceland, Ireland, Italy, Luxembourg, the Netherlands, Norway, Portugal, Spain, Sweden, Switzerland, Turkey, the United Kingdom and the United States. The following countries acceded subsequently to this Convention (the dates are those on which the instruments of accession were deposited): Japan (28th April, 1964), Finland (28th January, 1969), Australia (7th June, 1971) and New Zealand (29th May, 1973).

The Socialist Federal Republic of Yugoslavia takes part in certain work of the OECD (agreement of 28th October, 1961).

FOREWORD

The IEA Coal Industry Advisory Board (CIAB) commissioned its Committee on Coal Quality and Ash Characteristics to examine how boiler performance is affected by coal quality with full consideration of ash characteristics. The Committee has now completed its report with the contribution of many members of the CIAB. It is the outcome of a thorough enquiry carried out by coal producers and utilities with an OECD wide coverage.

The report represents the independent judgement of the CIAB. It is based on the experience of the practical problems encountered in coal use by large electric utilities located in different geographical areas of the OECD. It is a valuable contribution to determining whether the effects of coal quality on boiler performance were adequately defined and were receiving due attention in on-going research and current coal procurement practices. The CIAB believes that increased knowledge and understanding of this subject will assist the electric utilities of the IEA countries to achieve maximum cost effectiveness in the use of indigenous or imported coal to generate electricity.

I wish to thank the CIAB and all the experts who contributed their expertise to the successful completion of this report.

Helga Steeg
IEA Executive Director

I. INTRODUCTION

The Coal Industry Advisory Board (CIAB), at its meeting in April 1982, created a committee (shown on page 63) to study how boiler performance is affected by coal quality with full consideration of ash characteristics.

The Committee was chartered to investigate the magnitude of the relationship between boiler performance and coal quality and, if significant to define the economic impact based on current knowledge. Other aspects of the effect of coal quality (on a delivered basis to electric utilities) on the cost of electricity have been investigated by other committees of the CIAB, including studies regarding environmental control costs and transportation costs. The CIAB's Committee on Coal Quality and Ash Characteristics recognises a complete cost analysis of electricity as generated by a particular type of coal requires consideration of all cost aspects of the coal chain from mine to busbar. Equally important, or perhaps more so, the management practices associated with the operation of a particular power plant have impact on the availability of that power station and the cost of the electricity it generates.

However, when the CIAB set up the Committee on Coal Quality and Ash Characteristics, it especially sought to determine if the effects of coal quality on boiler performance were adequately defined and were receiving due attention in on-going research and current coal procurement practices. The CIAB believes increased knowledge and understanding of this subject will assist the IEA countries, and the electric utilities in those countries, to achieve maximum cost effectiveness in the use of coal to generate electricity. This is also true, but to a lesser degree, with respect to the use of coal in general industry. This Committee focussed its attention on the use of coal in electric utility boiler plants.

The Committee feels this issue is of substantial importance, especially to those countries increasingly dependent upon coal imports. As for any producer of electricity, the selection of coal by those entities that rely heavily on imported coal must be dictated by the economics of the total coal chain as measured by the additional need of coal importers for flexibility, diversity and predictability of coal supplies enhances the value of information regarding the relationships between coal characteristics versus boiler performances for this group of coal buyers.

Exhibit I
Imports Versus Use

OECD Europe	1982	1990	2000
Total Coal Use (Mtce)	336.1	392.1	488.1
. Imports (Mtce)	107.3	145.8	224.2
. Imports % of Use	31.9%	37.2%	50.0%

Note: Of the total of 112.0 Mtce forecasted increase in coal use, 108 Mtce (97%) is steam coal and the balance is coking coal.

Mtce = million metric tons of coal equivalent

Source: IEA/OECD Energy Balances and IEA Country Submissions (1983)

Of the OECD countries, the European countries and Japan will be most affected by these considerations. Exhibit I demonstrates that coal imported into Western Europe between 1982 and the year 2000 will increase from 32% to 50% of total coal use. Almost all of this increase will be for steam coal imports. Exhibits II and III provide details of these growth expectations.

Exhibit IV shows the recent history of world coal production. Traditional sources of thermal coal supply for the European countries and Japan (mainly domestic production) have all experienced negative growth in the last decade. Furthermore, the CIAB expects this pattern to continue through 2000 and beyond. Therefore, the European countries and Japan, of strategic and economic necessity, will seek to broaden their coal supply sources. To do so, coal buyers will need increasingly better evaluative tools to select and specify coal supplies.

Added to the issue of diversity and flexibility of coal supply, is the issue of rising fuel prices. Exhibit V shows the increase in steam coal prices (in national currencies) experienced in the United States, United Kingdom and Japan since 1973. While the rate of price escalation shown in Exhibit V will probably not continue at the same pace, this Committee believes the rise in fuel prices and the derivative rise in electricity cost to consumers have both had a negative effect on the economies of these countries. Efficient use of fuel is necessary to ensure that future increases in coal supply costs do not translate directly into increased electricity costs. We believe there is room for considerable improvement in boiler availability and efficiency throughout most of the IEA countries. A better understanding of the effects of coal quality and ash characteristics on boiler performance will help achieve that improvement.

Exhibit II
OECD Potential Coal Use by Sector
(Mtce)

	1982		1982		1982	
	Mtce	% Share [1]	Mtce	% Share [1]	Mtce	% Share [1]
ELECTRICITY GENERATION						
OECD Total	776.1	40.5	963.3	40.0	1307.9	42.4
North America	480.0	48.0	618.6	47.7	837.9	50.8
Pacific	52.2	19.2	83.7	25.0	136.1	28.9
Europe	243.9	37.3	261.0	34.2	333.9	35.1
Canada	24.9	18.3	30.0	15.2	46.4	13.1
United States	455.1	52.7	588.6	53.2	791.4	58.6
Japan	20.1	11.0	38.7	15.3	67.6	19.7
France	27.3	23.9	10.1	8.1	9.7	6.2
Germany	85.0	62.5	84.6	54.6	87.3	50.0
Italy	8.3	13.5	17.9	23.4	48.3	42.9
United Kingdom	70.3	71.8	62.9	66.2	65.7	58.7
INDUSTRY						
OECD Total	309.0	23.6	400.6	24.0	479.2	25.6
North America	152.4	24.0	199.1	22.9	233.6	24.6
Pacific	64.5	30.5	70.4	28.6	91.4	30.6
Europe	92.2	19.8	131.1	23.9	154.2	24.7
Canada	16.8	21.9	19.1	17.0	22.1	14.8
United States	135.5	24.3	180.0	23.7	211.4	26.4
Japan	51.6	29.3	53.3	27.1	72.3	29.9
France	11.9	16.7	19.1	30.6	22.4	31.8
Germany	24.7	24.7	26.6	24.5	26.9	22.4
Italy	7.5	12.7	10.0	14.9	13.3	16.9
United Kingdom	10.6	16.1	17.1	20.7	18.6	24.5

Source: IEA/OECD Energy Balances and IEA Country Submissions (1983)

1. For electricity generation, this represents the share of total electricity generated by solid fuels. For industry, this represents solid fuels share in total inputs.

The objectives of this report are:

(i) to define the present state of knowledge about the effects of coal quality and ash characteristics on boiler performance;

(ii) to identify areas where more knowledge is needed;

(iii) to make suggestions about how that can be done.

Exhibit III
OECD Projected Steam Coal Imports
(Mtce)

Steam Coal	1982	1990	2000	Rate of Growth (% per annum)
OECD Total	89.19	124.91	245.86	5.8
NORTH AMERICA	11.84	9.43	9.43	neg.
Canada	11.21	8.57	8.57	neg.
United States	0.63	0.86	0.86	-
PACIFIC	11.14	16.57	61.43	9.9
Australia	-	-	-	
New Zealand	-	-	-	
Japan	11.14	16.57	61.43	9.9
OECD EUROPE	66.20	98.91	175.00	5.5
Belgium	6.09	7.50	11.29	3.5
Denmark	8.27	10.43	9.00	0.5
France	13.51	16.29	30.43	4.6
Germany	10.27	10.79	10.79	0.3
Italy	8.87	19.57	54.11	10.6
Netherlands	5.41	7.71	8.86	2.8
Spain	2.73	4.70	11.26	8.2
Turkey	-	-	-	-
United Kingdom	1.20	1.43	-	-
Other Europe	9.84	20.50	39.27	8.0

Source: IEA/OECD Energy Balances and IEA Country Submissions (1983)

This report is not a technical treatise on the effect of coal quality and ash characteristics on boiler performance, but rather a guideline to that subject. Appendix B contains the bibliography developed by the Committee during the preparation of this report. Those seeking more detailed knowledge of the subject are encouraged to research the papers listed in the bibliography. The

Exhibit IV
Hard Coal Production by Regions/Countries [1]
(million metric tons)

	1973	1978	1981	1982	1983	Rate of Growth (% per annum)
TOTAL WORLD	2198.3	2552.6	2730.0	2819.5	2819.9	6.4
OECD [2]	912.9	944.4	1098.0	1107.0	1054.9	3.7
Canada	12.3	17.1	21.7	22.4	23.0	16.9
United States	529.6	576.8	700.8	707.2	660.0	5.7
Australia	55.5	71.8	87.4	91.1	99.5	15.7
Germany	103.7	90.1	95.5	96.3	89.2	neg.
United Kingdom	132.0	123.6	127.5	124.7	119.3	neg.
Other OECD	79.8	64.9	65.0	65.3	63.7	neg.
NON-OECD [3]	1285.4	1608.2	632.0	1712.5	1765.0	8.3
Africa	68.4	95.9	138.1	142.3	145.0	20.7
Botswana	-	0.3	0.4	0.4	0.4	-
South Africa	62.4	90.6	132.8	136.9	140.0	22.4
Zimbabwe	3.5	3.1	2.9	2.8	2.5	neg.
Other Africa	2.5	2.0	2.1	2.2	2.1	neg.
ASIA						
China	547.4	759.7	789.4	832.7	879.0	12.6
India	417.0	593.0	599.0	635.0	657.0	12.0
North Korea	30.0	35.0	36.0	36.5	45.8	11.2
South Korea	13.6	18.1	20.0	20.2	25.3	16.8
Other Asia	9.0	12.1	11.9	12.7	15.9	15.3
U.S.S.R	461.2	501.5	481.3	488.0	488.0	1.4
EAST EUROPE	196.7	233.0	202.6	227.6	229.0	3.9
Czechoslovakia	27.8	29.2	27.6	27.5	28.1	0.3
Poland	156.6	192.6	163.0	189.5	190.0	5.0
Other East Europe [4]	12.3	11.2	12.0	10.9	10.9	neg.
CENTRAL AND SOUTH AMERICA	11.6	18.0	20.5	21.8	24.0	19.9
Brazil	2.4	4.6	5.7	6.4	7.0	30.7
Colombia	3.0	4.9	5.0	5.6	6.2	19.9
Mexico	4.3	6.8	8.1	8.2	9.0	20.3
Other Central & S. America	1.9	1.7	1.7	1.6	1.8	neg.

1. Hard coal includes anthracite and bituminous coal
2. Source: IEA/OECD Coal Statistics
3. Source: United Nations Energy Statistics
4. Includes Bulgaria, East Germany, Hungary, Rumania and Yugoslavia

Note: 1983 figures are IEA estimates

Committee believes this bibliography is representative of the scientific information developed on the subject, although it does not claim to be exhaustive or complete.

Exhibit V
Fuel Prices for Electricity Generation
(in national currencies per toe)

	United States (in $/toe) Steam Coal	United Kingdom (in £/toe) Steam Coal	Japan (in Th. Yen/toe) Steam Coal
1973	15.80	11.10	5.35
1974	27.68	15.48	8.97
1975	31.75	23.60	11.18
1976	33.06	23.60	13.43
1977	37.08	32.76	16.08
1978	43.51	36.61	17.39
1979	47.73	42.20	17.98
1980	52.81	53.06	20.77
1981	58.78	62.38	23.70
1982	64.11	67.01	25.52
1983	64.30	71.58	n.a.

Note: toe = tonne of oil equivalent

Source: IEA/OECD Quarterly Price Statistics

In addition to researching the literature on this subject, the Committee conducted field research, interviewing electric utility personnel, boiler manufacturers and design engineers. Appendix A contains a list of those interviewed during the preparation of this report. In some cases the experiences discussed in this report are not attributed to any of the listed utilities or individuals interviewed. Many sources requested anonymity, not wishing their views to be publicly attributed. In the interest of researching the subject as completely as possible, the Committee agreed to maintain confidentiality and not publish the names of these interviewees.

II. ISSUES

The principal boiler performance issues studied by this Committee are as follows:

First, the Committee identified the major factors that affect boiler performance.

Second, the Committee surveyed the information and evidence developed regarding the economic impact of these factors on the cost of electricity.

Third, the Committee investigated how these factors are measured, defined and evaluated.

Fourth, the Committee looked at predictability. That is, given a certain set of physical characteristics, can the effect of a particular coal on boiler performance be predicted accurately without exhaustive testing?

The Committee documented on-going research into coal quality characteristics and their effects on boiler performance and how this research varies in philosophy and approach.

The principal problem affecting boiler availability and maintenance costs is tube failure. Thus, the principal issues investigated were slagging, fouling, corrosion and erosion. Similarly, the Committee examined those factors affecting boiler efficiency which generally led to tube failure, including carbon carryover and temperature distribution.

The physical characteristics of coal also affect the performance of peripheral equipment such as pulverisers and fans. While of considerable importance, these effects are not documented in this report to the same degree as slagging, fouling, corrosion and erosion. The Committee has concluded that the effects of coal quality on the peripherals is better understood by both coal users and producers, and therefore that it would not be useful to repeat that understanding herein.

III. CONCLUSIONS AND RECOMMENDATIONS

A. Conclusions

Based on the technical papers and operational results available, the CIAB's Committee must include:

1. The availability, capacity and cost of operation of each individual boiler are materially affected by the quality of coal fed to it. A summary of these effects is shown in Exhibit VI.

2. Coal quality is not only a function of consistency, but also a function of the various physical and chemical components of the coal and the coal ash.

3. The specific chemistry of the ash, which can give rise to varying degrees of slagging, corrosion and fouling in the boiler, has a major impact.

4. Available technical papers generally fail to address the economic impact on boiler performance of the phenomena they investigate. This makes relevance hard to establish and fails to get the attention of key decision-makers in the coal and utility industries.

5. The economics of the subject are gaining recognition among boiler manufacturers, electric utility companies and government agencies. There appears to be less recognition by coal companies, even those seeking new offshore markets.

6. The present methods that describe and specify coal quality require substantial refinement.

Exhibit VI
Coal Characteristics versus Effects

Coal Characteristics	Thermal Efficiency	Plant Availability	Operating Costs	Repair & Maintenance Costs
BY PROXIMATE & ULTIMATE ANALYSIS				
1. Moisture (Hygroscopic and Free)	X	X	X	X
2. Ash Quantity	X	X	X	X
3. Volatile Matter	X			
4. Calorific Value	X		X	
5. Sulphur	X	X	X	X
6. Hardgrove Index		X	X	X
7. Carbon	X		X	
8. Hydrogen	X			
9. Chlorine	X	X		X
10. Oxygen	X			
OTHER CHARACTERISTICS				
11. Coal Rank	X	X	X	X
12. Macerals (Petrographic Evaluation)	X	X		
13. Ignition Temperature	X	X		
14. Porosity, Sorption Potential	X	X		
15. Ash Viscosity	X	X		X
16. Initial Deformation Temperature	X	X		X
17. Softening Temperature	X	X		X
18. Fusion Temperature	X	X		X
19. Density and Friability				X
20. Burning profile (thermogravimetric analysis)	X	X	X	
ANALYSIS OF COAL ASH **Mineral Matter**				
21. Alumina Al_2O_3	X	X		X
22. Silica SiO_2	X	X		X
23. Titanium Oxide TiO_2	X	X		X
24. Ferric Oxide Fe_2O_3	X	X		
25. Lime CaO	X	X		
26. Potassium K_2O	X	X	X	X
27. Sodium Oxide Na_2O	X	X	X	X
28. Magnesium Oxide MgO				
29. Sulphur Trioxide SO_3		X	X	
30. Phosphorous Pentoxide P_2O_5				

7. Specific relationships between boiler performance and coal quality need to be developed. For example, the specific impact of sulphur, chlorine, sodium, overall ash content and coal rank (or reactivity) on carbon burn-out, slagging, fouling, corrosion and abrasion needs to be established. Until these relationships have been developed and proven, respecting differences in boiler design, coal buyers will continue to operate at a disadvantage when selecting new sources of coal and will be burdened by the need for exhaustive testing.

8. Economic parameters to measure the impact of boiler performance on the cost of electricity need to be established and agreed upon in the electric utility industry. For example, the cost of forced outages is presently disputed. Utilities with an excess of relatively inexpensive (at the margin) standby capacity view the subject one way. Utilities forced to turn to substantially more expensive sources of supply when base load units are out of service view the subject in an entirely different light. Generally, the capital cost implications of forced outages are ignored.

B. Recommendations

The CIAB Committee on Coal Quality and Ash Characteristics recommends that:

1. Both government and industry pursue the above questions with increased vigour. Specific issues to be pursued include:

    —   developing operating data that can provide a better correlation between the quality of coal actually used and the specific cause of forced outages;

    —   defining a methodology to cost forced outages in terms of maintenance, fuel substitution and capital cost;

    —   relating the specific aspects of coal quality to the frequency of outage, increased maintenance costs, reduced unit life expectancy and increased boiler heat rates;

    —   developing a universal method of defining coal characteristics so boiler performance can be predicted more effectively.

2. The IEA provides a forum to increase exchange of views and information about coal quality and boiler performance with an emphasis on the econmic relationships and advantages to be gained.

The Committee encourages the IEA to increase its use of that forum to exchange technical information, to encourage further research on the subject and to promote application of the existing knowledge in being to operating situations.

The Committee firmly believes that successful resolution of these issues is fundamental to achieving optimum use of an important energy resource — coal. Only by resolving these issues can we assure that coal will maintain its cost advantage over oil, thus enabling the IEA countries to reduce oil dependency in an economically advantageous way.

IV. COMBUSTION IN COAL-FIRED UNITS

Today, most of the coal used to generate electricity is consumed as pulverised fuel. Therefore, the focus of this report is on coal performance in pulverised fuel (PF) type utility units. This chapter briefly describes the combustion process in PF units and the coal-related performance problems these units may experience.

Steps in coal combustion:

- coal is pulversied in preparation for combustion;
- coal and air are injected into the combustion zone of the boiler;
- coal is ignited and burned in the combustion (radiant) zone of the boiler;
- combustion products exit the combustion zone and enter the convective sections of the boiler;
- fly ash and sulphur oxides (where flue gas desulphurisation is used) are removed from the combustion gases and collected;
- gases and other waste products exit the system.

The objective of the combustion process is simply to generate steam in the boiler to achieve design steam flow, pressure and temperature to drive the turbines that generate the electricity.

The key step of the process, in terms of boiler efficiency, is combustion. The combustion of coal has three phases:

- ignition;
- pyrolysis (combustion of the volatile matter);
- char burnout (combustion of devolatised coal).

In most utility units, combustion takes two seconds or less. During this short time, complex chemical reactions occur, mainly between various mineral components of coal, collectively referred to as ash content. These are explained in greater detail later in this chapter.

The performance of the unit is a function of two major elements:
- efficiency;
- availability.

Boiler efficiency is expressed as the amount of heat consumed to produce a given quantity of steam. Coal-related factors affecting boiler efficiency include:
- carbon carryover, coal that passes through the boiler with the ash, performing no useful work. Acceptable carbon carryover varies from 1% to 5% carbon in the ash;
- unburned coal volatiles (hydrocarbons) also leave the boiler without performing useful work;
- heat loss due to boiler slagging and fouling, throwing away heat instead of using it to generate steam;
- heat loss from exhaust gases, again losing heat that could have been used to generate steam.

The primary coal-related factors affecting boiler availability include:
- superheater tube failure due to corrosion, erosion, slagging or fouling;
- fireside wall damage due to slagging;
- temporary derating of a unit due to component failure (such as a pulveriser) or temperature imbalance, which can limit the amount of steam generated;
- wear and tear on forced and induced draft systems based on ash loading and abrasion.

Thus, boiler availability can be measured either as a function of forced outage, during which time the boiler is 100% unavailable, or as a function of partial outage, during which time the boiler operates below design capacity.

The most troublesome coal-related problems in the boiler are:

- — slagging;
- — fouling;
- — corrosion and abrasion;
- — temperature imbalance.

The first three are primarily caused by coal ash and its characteristics. The fourth, temperature imbalance, can be caused by coal ash, by the characteristics of the coal itself, or by a combination of the two.

Exhibit VII illustrates a typical boiler.

Boiler slagging is commonly defined as deposits of mineral matter that fuse and form on furnace walls and other surfaces in the radiant heat (combustion) zone of the boiler.

If slagging does occur it can creep into the convective sections of the boiler. Heat transfer in the radiant zone is impeded by slag formations and raises the temperature of gases entering the convection section. Heat is absorbed in the radiant section as water is converted to steam in the tubes lining that section. The rate of heat absorption in the radiant section is important to ensure the gases are sufficiently cooled before entering the convective section.

Slagging in the boiler reduces efficiency and causes damage. Slagging reduces efficiency by impeding the transfer of heat to water to generate steam. As slag deposits increase, the ability to generate steam is reduced and partial derating of the boiler can occur. Ultimately, accumulated slag deposits can fall to the bottom of the boiler. These pieces of slag can be as large as 2 or 3 tons in a large unit and significantly damage the radiant section of the boiler when they fall. Such damage is likely to cause forced outage and reduce unit availability. Slagging also causes temperature imbalance in the boiler since combusted gases in the radiant section do not cool. Instead, excessively hot gases enter the convective section and cause superheater temperatures to run out of control, endangering the unit and possibly causing tube failure.

Fouling refers to high temperature, bonded deposits that form on the superheater and reheater tubes in the convective section of the boiler. Convective section fouling impedes, and can block, exhaust gases passing through the convective section. Fouling also impedes transfer of heat through superheater tube walls, thus partially derating the boiler.

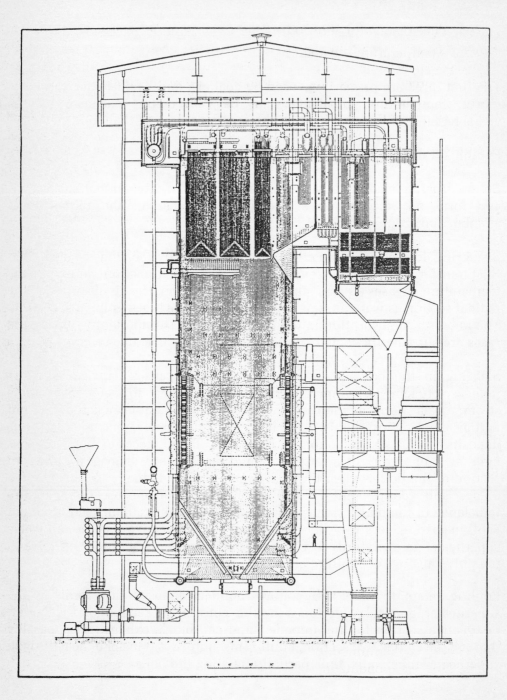

Damage to superheater tubes can occur when mineral matter from coal ash is deposited on the tubes (fouling) and is corrosive. In this way, corrosion and fouling are linked.

Abrasion occurs when coal ash particles, travelling at high speed, impinge on boiler surfaces. Depending on the amount and abrasive quality of the ash, abrasion can become a serious problem causing reduced unit availability either by forced outage or increased maintenance schedules.

Temperature balance in a boiler is also affected by the burning profile of the coal: the rate at which coal passes through the three combustion steps, and at what temperature. While the burning profile is not entirely independent of ash content, the volatile matter content and maceral types in the coal have greater impact.

During combustion, temperatures near 1800 °C, but the gases must cool to about 1200 °C before entering the convective sections of the boiler. If the coal burns too quickly or too slowly the temperature requirement will not be met.

If the coal burns too quickly:

- too much heat can be absorbed in the radiant section of the boiler. When escaping gases reach the superheater tubes they are too cold to raise steam temperature to the levels necessary for efficient turbine operation and full-capacity utilisation;

- temperatures in the radiant section can rise too high and cause circulation problems or increased boiler slagging, thus raising the incidence of forced outages.

If coal burns too slowly, temperatures in the radiant section do not reach design levels, and gases reaching the superheater tubes may be hotter than 1200 °C. Thus, there can be a decrease in boiler efficiency through:

- decreased steam production;

- fouling of superheater tubes;

- increased carbon carryover;

- loss of superheater temperature control;

- higher than desired exit temperature of exhaust gases.

As previously mentioned, a series of complex chemical reactions involving the mineral components in coal ash, takes place. Exhibit VIII is a simplified

Exhibit VIII
Effect of Heating on Minerals in Coal

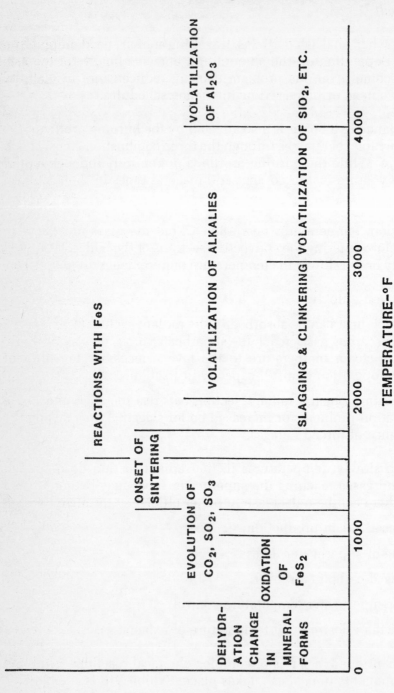

EFFECT OF HEATING ON MINERALS IN COAL

Source: Combustion Engineering Inc.

representation of these reactions, which are extremely important because they affect the formation of chemical compounds that promote slagging, fouling and corrosion.

The most important chemical reactions are those affecting sodium and potassium, the alkali metals. The interaction of these metals with chlorine, sulphur and other ash components is critical. Fouling and corrosion result mainly from selective condensation of alkali metal salts on superheater tubes. These salts are extremely corrosive and the principal cause of damage to superheater tubes.

Finally, coal has a significant effect on boiler design and in turn on the capital costs of a utility generating facility. Coal quality directly affects plant design. Coals with difficult characteristics can be compensated for, but only at high cost. For example, boilers designed to operate with coals possessing slagging and fouling tendencies are larger than units operating with coals with minimal tendencies to slag or foul. Thus, the expected slagging and fouling tendencies of the coals used are a major cost consideration for any unit. Likewise, coal quality considerations affect the cost of peripheral equipment associated with the unit. For example, particularly hard coals or those with particularly abrasive mineral content require a more expensive and higher capacity pulveriser installation than less demanding coals.

Finally, capital costs, as amortized over kilowatt hours generated, can rise as boiler availability decreases. Also difficult coals can accelerate the physical depreciation of plant and equipment.

V. RESULTS OF THE FIELD INVESTIGATION

Discussions with coal procurement and boiler operating personnel were conducted by this Committee in a number of countries, including:

- Australia;
- Denmark;
- Holland;
- Italy;
- Japan;
- Republic of South Africa;
- United Kingdom;
- United States.

The overall sentiment expressed by utilities during these interviews was one of frustration. They strongly feel the lack of evaluative tools to decide which buyable coals will cause boiler problems and which coals will not. Examples were cited showing coals of similar characteristics to perform quite differently in different boilers. Coals predicted to perform well, performed poorly and vice versa. The consensus is that the evaluative tools are inadequate and inaccurate. There is still no substitute for substantial testing of coals. This testing is time-consuming and expensive.

This has been a particularly difficult problem for utilities seeking to change fuels in existing units. There is less of a problem when designing new boiler plants to fit a particular coal or range of particular coals. However, this area is not free of difficulties. Examples were cited where new boiler plants were designed to operate with a particular coal and yet experienced substantial slagging and fouling problems with those same coals.

In most of the countries included in the field interviews, coal-fired power stations are base loaded and high in the merit order of dispatch. Therefore, lack of availability through total or partial forced outage results not only in expensive maintenance, but also in expensive fuel substitution costs. In most cases, the expensive substitute fuel is imported fuel oil. In other countries, however, there is spare coal-fired capacity at present so substitute generating costs are not significantly higher than the costs of the units forced out of service.

Specific comments encountered during the field interviews:

– inexplicable, and often irreconcilable, differences in coal sampling have been experienced between samples taken at loading port and port of discharge. These differences are often significant and have negative effects on boiler performance;

– while it is possible to design a boiler to burn a wide range of coals, "if you design for everything you cannot afford the boiler";

– "design of turbines is a science; design of boilers remains an art";

– distribution of coal particle sizes depends to a large degree upon physical characteristics of the coal and affects the performance of the distribution system between the pulverisers and the burners. It is often difficult to predict problems in this area, but they can seriously affect boiler performance;

– different coals with apparently similar calorific and volatile matter content produce unexplained and markedly different termperature distributions throughout the boiler to the disadvantage of boiler efficiency and final steam temperature;

– low sulphur coals tend to cause problems in the precipitators;

– some coals have low density ashes that cause problems in precipitators;

– the impact of chlorine is not well understood;

– there appear to be no internationally accepted benchmarks for satisfactory generating unit availability. The National Electricity Reliability Council (NERC) of the United States during the period 1971-1980 cites average availability in the United States as 68.8% for coal-fired units of 400 MW and above;

  . some utilities interviewed indicated based-loaded, coal-fired units were operating at much higher levels of availability, while others were much lower;

. forced outage of 7% or more is generally considered unaccept-
able;

. a target of 2% maximum for forced outages was cited by some as
desirable;

— based on the experience of one Australian utility it was estimated
that the cost of forced outages and reduced boiler life due to
inconsistent coal ash levels was A$10.60 per metric ton of coal
(based on normal annual coal consumption levels);

— in another Australian utility plant, a 5% forced outage rate resulted
in an estimated cost to the system of A$ 11.18 per metric ton of
coal;

— in both Australian instances, a solution to the problem was
approached by installing coal washeries to reduce ash content and
(most important) create consistency in ash levels in the coal (see
Chapter VIII);

— in general, utilities do not keep data comparing boiler availability
or efficiency with type or source of coal used. Therefore, it is
difficult to determine specific relationships between boiler
performance and coal supplies in historic context.

While most utility personnel interviewed, particularly those using substantial amounts of imported coal, agreed that efficient and effective operation of
their boiler plants is sensitive to coal characteristics, there was concern that
coal suppliers do not appreciate all the economic factors involved in
selecting coals. For example, balance of trade considerations are important to
some countries that import coal. In those cases, the governments may accept
higher than normal maintenance costs (for example), which are domestic
currency costs, to save foreign exchange. Thus, most utilities that import
substantial quantities of coal emphasized a need on the part of the CIAB, the
IEA and other industry observers to recognise the complex questions facing
utilities when procuring coals overseas.

VI. PRESENT STATE OF RESEARCH

A. General

Most of the research identified by this Committee focussed on the phenomena that led to slagging, fouling and abrasion and the chemical reactions that occur during combustion. More recently, research has been conducted to examine the effects of the maceral contents of the coal on the combustion process. These two broad areas are discussed in this chapter.

The basic thrust of the research to date has been to develop a better understanding of the phenomena that occur and to improve methods for measuring coal quality characteristics as they relate to these phenomena so performance can be reliably predicted.

The most active centres of research activity appear to include:

- United States;
- United Kingdom;
- Canada;
- Australia;
- Japan;
- West Germany.

The operation or applications oriented research that was identified emphasized improvement in boiler plant design to fit particular coals or ranges of coals. Little emphasis was placed on the question of the impact of differing coal qualities on existing units, particularly coals whose characteristics fall outside the design parameters for existing units.

Furthermore, very little research has examined some of the questions uncovered during the Committee's field survey. For example, explaining why coals that appear quite similar perform quite differently in a given boiler. Finally, there appears to be more interest in the research to explain phenomena and less interest to determine the operational and economic impact of these phenomena on boiler performance. There is even less effort in the research to extend analysis to the impact on the cost of electricity.

From the viewpoint of the operating utility executive, or the executive charged with selecting and procuring coals for existing power plants, the research remains more academic than practical.

There are some exceptions to the lack of economic focus. Recent work in the United States (notably at TVA, EPRI and Amercian Electric Power) has started to examine the impact of changes in coal quality on the cost of electricity. Likewise, the CEGB in the United Kingdom has also started to examine this issue, although no public report has yet emerged.

In Germany, the research emphasizes new coal utilisation technologies. The Committee found little in the German literature documenting the affects of coal quality on boiler performance.

In South Africa and Australia emphasis is on developing boiler design criteria to enable use of lower-quality indigenous coals.

The following sections describe the evident state of knowledge about coal quality, coal ash and ash chemistry as revealed by the literature. The principal aspects discussed include:
- ash quantity;
- ash chemistry;
- sulphur;
- chlorine;
- coal macerals and rank.

B.   Ash Quantity

Assessing the effect of ash quantity without reference to ash chemistry is not entirely useful. It is clear from the literature that ash quantity by itself has limited impact on boiler performance, principally in:

- reduction of boiler efficiency due to loss of sensible heat in ashes (small effect);

- added wear and tear on pulverisers;

- increased abrasion of boiler components.

The affect of varying ash quantities is especially small if that quantity is within the boiler design parameters. Some evidence developed in the United States indicates that deviation on either side of design ash quantity specifications can produce a negative impact on boiler performance (40).

Other experiments in the United States, however, demonstrate that a reduction in the ash content of troublesome coals (those with high slagging and fouling tendencies) substantially reduces slagging and fouling problems. Thus, it appears that ash quality characteristics are more important than ash quantity characteristics when predicting boiler operating problems with a particular coal. Ash quantity is critical with troublesome coals.

Nonetheless, the Committee wishes to emphasize that, when examining all the cost factors (mine to busbar) associated with the impact of coal costs on electricity generated, total ash content can be economically important whether or not the coal is troublesome. This is especially true when coal must be transported long distances from mine to electricity generating plant. The effects of ash quantity on boiler peripherals (pulverisers and fans especially) are important and cannot be ignored.

"From the user's standpoint, there is an indisputable case for reduction of the absolute minimum ash content in coal delivered to pulverised coal-fired boiler plants. It is a curious twist of the economic structure that causes the ash to be transported over considerable distances from the pit to the power station where, far from serving a useful purpose, it causes trouble in all directions. The case is made more curious when, after causing so much difficulty in the boiler plant, a considerable amount of money and effort has to be spent in collecting and removing the ash to a convenient dumping ground" (10).

C. Ash Chemistry

As previously noted, chemical reactions that occur in the combustion zone of the boiler are complex and rapid. Elevated temperatures and high speed make them difficult to observe. One of the principal problems in identifying the chemical reactions and their affect on boiler performance has been to identify the intermediate chemical compounds produced.

"Despite the extensive and numerous studies of ash deposition and corrosion which have been made in the past, laboratory tests capable of revealing precisely how a previously untried coal will behave in a boiler have yet to be devised" (13).

That quote is from a report prepared by the British Coal Utilisation Research Association, Boiler Availability Department, in 1967, yet the point still remains valid. Typically the ash chemistry of a particular coal is determined by preparing a laboratory ash analysis. This analysis involves burning a coal sample under controlled conditions and chemically analysing the composition of remaining ash. This is done in the United States according to the ASTM D 3174 ashing procedure. This method of ash analysis typically shows about ten oxides of various minerals. Examples include silicon dioxide, alumina, titanium oxide and ferric oxide.

"Since these measurements are based on laboratory-made ash prepared at around 800 °C, a temperature which destroys many of the mineral phases present in the original coal, the tests are performed therefore on the decomposition products of the mineral matter rather than on the original mineral matter associated with that coal. In addition, the laboratory test cannot directly simulate the conditions within a large PF furnace where gas-borne particles are subjected to very rapid heating and cooling" (18).

Another commonly performed test addresses the ash fusion temperatures of coal ashes. In this test coal ashes are heated under controlled conditions and the points at which the ash initially becomes soft, deforms and becomes fluid are observed.

From these two tests, formulas can be applied to estimate the slagging and fouling potential of a coal.

The most common formula derived from laboratory ashes is the base-to-acid ratio. From this ratio the slagging and fouling potential of a coal can be estimated. However, application of the base-to-acid ratio formula is limited to certain kinds of coals. It does not apply to coals with lignitic type ashes (the sum of calcium and magnesium oxide exceeds the content of ferric oxide) (26).

Slag viscosity has been used in both the United States and Australia to predict slagging propensity. The research indicates this method is more accurate and more universally applicable to different coal ash types than the base to acid ratio method (24)(26).

Slag viscosity can either be determined by laboratory measurement or calculated from ash analysis. The calculated viscosity can be less time-consuming and expensive than the laboratory measurement of slag viscosity.

The most troublesome ash components are sodium, potassium and phosphorous (chlorine, also a troublesome coal component, is not a component of the coal ash). Some researchers believe the soluble sodium content of laboratory ash is a direct and reliable indicator of the amount of sodium in the coal and therefore an indicator of its propensity for fouling and corrosion (13).

Another index of fouling and corrosion propensity is the sodium-to-ash ratio. Some studies have shown that fouling becomes severe for British coals when this ratio exceeds 0.010. However, sodium cannot be considered independently of chlorine content. There is some evidence that the chlorine-to-total sodium ratio varies in direct proporation to the propensity of a coal to cause corrosion and fouling problems. When this ratio exceeds 2:1 some researchers believe a coal may become a problem (13).

Use of ash fusion test data can be misleading. Ash fusion tests typically are run in both a reducing and oxidising environment. This means there is either sufficient oxygen in the atmosphere surrounding the ash particles to oxidise various minerals or there is not. Generally, an oxidising environment pertains throughout the combustion chamber of the boiler. However for a number of reasons there may moments when, as the coal and mineral particles pass through the combustion chamber there is not enough oxygen for oxidation to occur. This is known as a reducing atmosphere. The ash fusion temperatures are somewhat lower in a reducing environment. It is important to be aware of these temperatures since, if a reducing environment develops, the ash fusion temperatures can become low enough to cause slagging or fouling.

The problem with ash fusion measurement is that hot stage microscope studies show that two ashes with similar fusion characteristics have markedly different melting and crystallisation behaviour (18), and thus could perform very differently in the boiler. In such a case the ash fusion test would not function properly as a predictor of performance.

In short, using ash fusion temperatures alone to predict coal fouling and slagging tendencies can be inconclusive at best.

Despite the availability of all this testing methodology the Committee's field research shows that utilities have had disappointing results with these predictive tools. The general sentiment is that better tools are needed.

As noted earlier, the problem with ash analysis and the prediction of fouling and corrosion tendencies based on ash analysis is that a number of complex chemical reactions occur in a chain as temperatures change during the short time that particles are in the combustion zone. The degree to which these chemical reactions occur and the speed at which they move from one reaction to another as the various elements associate and disassociate, not the simple presence of certain chemical elements, determines if there is potential for slagging, fouling and corrosion. The balances are very delicate and that makes accurate prediction of the occurrence of these reactions extremely difficult. Simply viewing « the decomposition products » of these minerals is not enough.

Two new techniques, depositional testing and preparation of "low temperature ashes" are being developed to address this problem.

Depositional tests are simply the combustion of coals in pilot size boilers under very carefully controlled conditions where the superheater and combustion regions of the boiler are simulated. Coals are burned in these pilot units much as they would be burned in a full-scale boiler. The deposition of fouling and corrosive deposits is assessed and examined in these units. Use of these units is reported in Canada, the United States and the United Kingdom for predicting coal performance in boilers.

"Low temperature ashes" is a method of separating the coal from the ash-producing mineral matter without oxidising the minerals (18). After separation, experiments are performed on those minerals to examine changes of state as they occur to predict slagging, fouling and corrosive properties.

D.   Sulphur

The effects of sulphur on boiler performance are difficult to categorise as good or bad.

Sulphur contributes to low temperature corrosion in a utility power plant. Sulphur is also associated with certain environmental problems.

However, reducing sulphur below certain levels has a negative impact on the operation of precipitators. There is also evidence in some of the research that, especially for coals with high chlorine content, higher sulphur levels may reduce high temperature corrosion and fouling.

On the other hand, with some coals slagging potential increases as sulphur content increases (26).

E.   Chlorine

In many ways, chlorine is the most troublesome component of steam coal. There is substantial evidence that fouling and corrosion increase as the chlorine content in coal increases (13). The chlorine combines with either sodium or potassium to form alkali metal salts. These salts are the primary cause of fouling and corrosion.

Some researchers believe the chlorine-to-ash ratio can be used to predict corrosion and fouling. An increase in ash content relative to chlorine reduces fouling potential (13).

As noted above, there is also some evidence indicating the sulphur-to-chlorine ratio is important. As that ratio increases, coals become less corrosive (13).

F.   Petrographics

Petrographics is the science of analysing the components of the coal itself. Coal rank is most accurately determined by petrographic analysis. While volatile matter is a more common measurement of coal rank, it is not as accurate as petrographic examination.

Since coal rank is the primary determinant of coal reactivity and burning profile, petrographic analysis is a tool for predicting the coal burning profile and reactivity. The main petrographic analytical technique is measurement of the reflectance of the vitrinite in coal (17)(19).

An increase in the vitrinite reflectance relates directly to an increase in coal rank. As a general rule, as coal rank increases, reactivity decreases. More specifically, reactivity of a steam coal is determined by measuring the relative quantities present in the coal of the three principal maceral groups.

- vitrinite;
- exinite;
- inertinite.

Exinite contains the most volatiles and is the most reactive. Following exinite is vitrinite. Inertinite is the least reactive. The higher the inertinite content, the lower the reactivity and combustion efficiency. Conversely, the higher the exinite, the higher the combustion efficiency. Vitrinite adds to combustion efficiency but not to the same degree that exinite does (17).

It appears that this kind of petrographic classification will be useful to predict burning profiles. Thus, it has the potential to be useful in selecting untried coals for existing units.

Many other measurements of reactivity are used at present. For example, the Japanese use a Fuel Ratio (fixed carbon divided by volatile matter). They find this ratio useful for predicting reactivity, carbon carryover and formation of the oxides of nitrogen.

G.   Nitrogen

The nitrogen content in coal has little or no known effect on boiler performance. However, research in Japan indicates that the nitrogen content in the coal has a significant impact on the generation of nitrogen oxides during the combustion process. Thus, for environmental cost control reasons, nitrogen content becomes important.

H.   Moisture

Moisture content in coal has an effect on boiler and system performance. In the boiler, moisture removes sensible heat for the boiler that is largely lost for the purposes of generating steam. The Committee finds that an increase of 5% in total moisture results in a decrease of 0.3% in boiler efficiency (assuming initial moisture content of about 10% or more).

Outside the boiler moisture content is also important to control. Too little moisture (the actual level depends on the rank and other characteristics of the particular coal involved) will permit dust to become a problem when handling coal.

On the other hand, too much moisture can cause freezing problems when handling coal in cold weather. Moisture content also promotes spontaneous combustion problems with some coals. Some coals, when too moist, will clog coal hoppers and feeders.

Finally, excess moisture causes problems in pulverisers.

In general, the desirable range for moisture content in coal lies between 5% and 10%.

VII. ECONOMICS

As previously noted, the economic impact on the cost of electricity generated due to coal quality and its effects on boiler performance are not well established. Recent attention has been directed to this question in the United States and the United Kingdom. Specific reference is made to the work of the Electric Power Research Institute and the Tennessee Valley Authority (TVA) in the United States and to the Central Electricity Generating Board in the United Kingdom. Some investigation of the question has been conducted in Australia.

A.   United States Experience

The principal methodology that has been used in the United States to examine the economic impact of coal quality is regression analysis using historic data (35). However, as pointed out in the TVA study, this approach suffers from the lack of correlative detail available with respect to historic boiler performance and fuels used. Researchers have relied mainly on coal purchase documents and assumptions about when various types of coal were taken from stocks to be used in boiler plants.

One of the principal recommendations emerging from the TVA study (35) is that utilities begin to track closely the quality of coal being used on a daily basis. A second recommendation was to be more specific in reporting boiler outages to define the problem that forced the outage. In the past, descriptions of what the actual forced outage problem was have been too generalised to enable accurate correlation of coal quality to specific boiler operating problems.

The CIAB Committee on Coal Quality and Ash Characteristics endorses those recommendations.

The TVA report warns that the studies are unique to the TVA system. Specifically, it says, "The data may be biased by TVA's coal-purchasing policies, operating practices or boiler maintenance philosophy. However, this study documents where the utility industry is in regard to coal/plant performance relationships." (35).

With respect to boiler efficiency the report goes on to state: "An analysis of data from most TVA units (omitting Allen, Cumberland and Watts Bar) showed that coal ash and moisture and boiler age had the greatest effect on boiler efficiency. The following relationship was derived for boiler efficiency for the TVA system as a whole:

boiler efficiency = K- 0.022 (ash) - 0.100 (moisture) -0.039 (age).

The constant K has a different value for each group of like boilers. Coal sulphur and boiler load were investigated as possible variables but were not shown to be important statistically.

"The above model accounted for 91% of the variability in reported efficiency values."

Regarding boiler availability, the study determined the following relationship:

1n (unplanned outage hours) = K + 0.032 (ash) + 0.293 (sulphur) + 2.176 [(1n 245 + arc tan) age - 18.8]/8.1).

Regarding ash, the study points out: "Over the range of ash fired at TVA plants (generally 12% to 24%), the relationship indicated that, for a typical plant, the outage hours can vary by 360 hours per year because of changes in ash alone." Regarding the effects of sulphur, the study points out: "Over the range of sulphur values for TVA plants (generally 1.0% to 5.0%), outage hours at a typical plant may vary by as much as 870 hours per year due to sulphur alone." (35).

With respect to maintenance costs, the study determined the following relationship:

1n (maintenance cost) $= K + K_x$ (ash lagged six months) $+ K_y$ (sulphur lagged ten months).

The correlation co-efficient was equal to or exceeded the 90% confidence level.

B.   Field Interview Experience Regarding Economics

During the field studies conducted by the Committee, some economic penalties associated with ash content were reported. It appears that an increase of 1% ash (generally after passing the 10% ash level) results in a decrease of about 1.2% to 1.5% in boiler availability. Assuming capacity costs about $1,000 KW, the capital cost absorption penalty is equivalent to about $0.95 per ton of coal burned per 1% increase in ash content in the coal.

Likewise, a 1% increase in ash (again over the 10% range) results in a decrease of about 0.3% in boiler efficiency. Based on the field interviews, this costs about $0.67 per ton of coal burned.

Taken together, these two factors can result in a cost penalty of about $1.62 per ton of coal per 1% increase in ash content, disregarding all but boiler efficiency and availability factors.

No similar economic relationships between ash chemistry (as opposed to ash quantity) and boiler performance were found to have been developed, even though these relationships are recognised as being more significant.

C.   Australian Experience

The experience in Australia where power utilities are government owned, with problems of boiler availability due to changes in ash content and character show larger numbers. The Committee cautions that these results may represent an exaggerated case.

In 1982, the Ombudsman of New South Wales, Australia, conducted a public inquiry into power station failures. During the 28-month period from 31st October 1979 to the time of the inquiry 59 boiler tube failures were reported, causing a lost output of 2,014 GWh, equivalent to an average forced outage rate of 7.2%. The inquiry determined that this forced outage rate was due primarily to ash content exceeding design levels and to inconsistency of coal quality.

The inquiry concluded that the reduced output resulted in increased fixed capital charges, operating and maintenance costs equivalent to an increased cost of $A6.80 per metric ton of coal consumed. This cost did not include replacement power cost. Furthermore, the shortened life of major boiler components was determined to increase power production costs by an additional $A4.20 per metric ton. Thus, the overall cost of the forced outages and reduced life expectancy was calculated to be $A10.60 per metric ton of coal consumed. As previously noted, the problem was addressed by construction of coal washeries to reduce ash content and introduce more consistency in the quality of the coal reaching the power plants.

D.   Japanese Experience

The experience in Japan seems to confirm the technical evidence regarding emphasis on consistency of coal quality and the need to have delivered coal qualities remain within the design parameters of the boiler. Exhibit IX shows that the experience of the Japanese electric utilities with respect to forced outages has been exceptionally good, with forced outage rates generally less than 2% annually. The exhibit also shows that the quality of coal consumed has been consistent. They report that the ash levels are according to boiler design specifications.

However, the Japanese also report that they emphasize heavily preventive maintenance and training of operations. Boilers are inspected annually. Thus, some of the repairs that might be made elsewhere during periods of forced outage may be done in Japan during planned preventive maintenance periods. Those periods of plant inspection and shutdown are mandated by Japanese regulations. They range from 20 to 70 days or more depending on unit size and overhaul requirements.

Thus, there is an economic trade-off between forced outage versus planned maintenance and inspection hours. The Japanese appear to have elected higher planned outage time in order to reduce forced outages to very low levels.

Exhibit IX

1. Forced Outage Rate versus Coal Quality, Japan

	1967	1968	1969	1970	1971	1972	1973	1974	1975	1976	1977	1978	1979	1980	1981
Number of Units [1,2]	38	42	44	44	37	31	19	15	14	14	12	12	12	14	16
Forced Outage Rate [3,4] (%)	1.7	1.2	1.9	1.2	1.3	5.0	7.0	1.4	1.2	0.9	0.6	0.3	1.0	1.7	1.9
Plant Factor [5] (%)	78.0	76.3	75.2	73.4	64.9	59.9	66.9	68.4	65.9	69.2	69.9	67.8	72.7	72.5	74.2
Average Generating Days	321	312	320	330	322	298	301	312	294	302	306	301	304	286	290
(Coal Quality)															
H.C.V. [6] (kcal/kg)	5840	5740	5670	5590	5540	5660	5570	5460	5580	5760	5840	5900	5910	5980	6200
Ash Content [6] (%)	24.7	26.1	26.9	27.8	27.9	26.4	26.6	28.3	27.3	25.3	24.5	24.2	24.4	22.0	20.0

1. Excludes plants of Joban Joint Venture Power Company
2. Excludes units less than 156 MW of unit capacity
3. $\dfrac{\text{Forced outage days}}{\text{Service days + Forced outage days}} \times 100$
4. Includes forced outage caused by grid operation failure outside the power plants
5. $\dfrac{\text{Output at kWhs}}{\text{Nameplate rating at kWs} \times \text{Period hours}} \times 100$
6. Air dried basis

Source: Electric Power Development Company

Exhibit IX
2. Plant Performance versus Coal Quality, Selected Units

Isogo Power Plant Unit #1*		1974	1975	1976	1977	1978	1979	1980	1981
(Plant Performance)									
Availability [1]	(%)	86.3	86.9	86.3	91.4	86.6	91.0	85.9	92.1
Plant Factor [2]	(%)	67.3	71.8	73.8	80.3	73.9	81.1	79.9	84.3
Thermal Efficiency [3]	(%)	39.4	39.5	39.5	39.5	39.6	39.5	39.4	39.3
Forced Outage Rate [4]	(%)	1.0	1.3	1.3	0.1	0.9	1.5	1.7	1.2
(Coal Quality)									
H.C.V. [5]	(kcal/kg)	6110	6210	6270	6300	6380	6350	6410	6440
Ash Content	(%)	16.8	16.9	16.0	15.4	15.7	16.4	17.1	17.4

* Commissioned by EPDC in May 1967 with 265 MW of capacity

Exhibit IX
3. Plant Performance versus Coal Quality, Selected Units
(continued)

Isogo Power Plant Unit #2*		1974	1975	1976	1977	1978	1979	1980	1981
(Plant Performance)									
Availability [1]	(%)	91.4	80.3	91.0	86.6	92.6	87.8	90.1	87.1
Plant Factor [2]	(%)	79.6	68.8	77.2	75.3	78.6	79.2	82.6	78.9
Thermal Efficiency [3]	(%)	39.5	39.5	39.5	39.5	39.6	39.4	39.3	39.2
Forced Outage Rate [4]	(%)	0.4	2.6	0.6	0.3	0	2.6	1.5	0.6
(Coal Quality)									
H.C.V. [5]	(kcal/kg)	6090	6190	6270	6300	6390	6350	6380	6430
Ash Content	(%)	17.4	17.1	16.1	15.8	15.9	16.5	17.0	17.5

* Commissioned by EPDC in September 1969 with 265 MW of capacity

Exhibit IX
4. Plant Performance versus Coal Quality, Selected Units
(continued)

Takasago Power Plant Unit #1*		1974	1975	1976	1977	1978	1979	1980	1981
(Plant Performance)									
Availability [1]	(%)	86.8	90.7	84.9	94.2	84.0	89.8	87.0	92.7
Plant Factor [2]	(%)	66.7	79.3	72.3	81.6	75.3	77.8	75.3	81.4
Thermal Efficiency [3]	(%)	37.5	37.8	38.1	37.1	37.7	37.7	37.6	37.6
Forced Outage Rate [4]	(%)	3.2	3.2	0.6	1.4	0.2	2.8	1.4	0.4
(Coal Quality)									
H.C.V. [5]	(kcal/kg)	5900	6290	6310	6350	6380	6380	6340	6440
Ash Content	(%)	25.0	21.8	20.9	20.6	20.2	21.0	21.3	20.7

* Commissioned by EPDC in July 1968 with 250 MW of capacity

Exhibit IX
5. Plant Performance versus Coal Quality, Selected Units
(continued)

Takasago Power Plant Unit #2*		1974	1975	1976	1977	1978	1979	1980	1981
(Plant Performance)									
Availability [1]	(%)	89.6	86.2	91.2	86.0	92.3	90.6	90.5	79.6
Plant Factor [2]	(%)	67.5	72.4	78.1	74.9	78.6	77.0	78.0	71.5
Thermal Efficiency [3]	(%)	37.5	38.0	37.8	37.3	37.7	37.6	37.6	37.6
Forced Outage Rate [4]	(%)	2.1	1.4	0.6	0	0.3	0.2	3.1	6.0
(Coal Quality)									
H.C.V. [5]	(kcal/kg)	5850	6130	6300	6330	6390	6410	6330	6430
Ash Content	(%)	25.3	22.9	21.3	20.8	20.5	20.4	21.2	20.6

* Commissioned by EPDC in January 1969 with 250 MW of capacity

— 47 —

Exhibit IX
6. Plant Performance versus Coal Quality, Selected Units
(continued)

Takehara Power Plant Unit #1*		1974	1975	1976	1977	1978	1979	1980	1981
(Plant Performance)									
Availability [1]	(%)	84.5	90.7	84.0	89.2	81.5	90.2	91.8	78.6
Plant Factor [2]	(%)	69.3	75.1	76.8	75.0	72.1	87.2	88.0	77.3
Thermal Efficiency [3]	(%)	37.9	37.9	37.9	37.7	38.0	37.8	37.2	37.2
Forced Outage Rate [4]	(%)	0.8	0.3	1.1	1.7	0.5	0.6	0.3	3.7
(Coal Quality)									
H.C.V. [5]	(kcal/kg)	5960	6140	6140	6290	6310	6310	6270	6200
Ash Content	(%)	23.4	21.8	22.3	21.2	21.5	21.3	21.7	22.1

* Commissioned by EPDC in July 1967 with 250 MW of capacity

Exhibit IX
7. Plant Performance versus Coal Quality, Selected Units
(continued)

Matsushima Power Plant Unit #1*		1980	1981
(Plant Performance)			
Availability [1]	(%)	100.00**	80.3
Plant Factor [2]	(%)	94.9	77.8
Thermal Efficiency [3]	(%)		39.1
Forced Outage Rate [4]	(%)	0	0.7
(Coal Quality)			
H.C.V. [5]	(kcal/kg)	6,560	6,530
Ash Content [5]	(%)	14.4	14.4

* Commissioned by EPDC in January 1981 to burn imported coal, with 500 MW of capacity

** With the start of operation in January 1981, this unit did not enter into scheduled shutdown for periodical unit inspection in this period (fiscal year 1980), and there was no limitation on unit availability by the unit inspection/overhaul which is required by law every one year operation

1. $\dfrac{\text{Available days}}{\text{Period days}} \times 100$

2. $\dfrac{\text{Output at kWhs}}{\text{Nameplate rating at kWs} \times \text{Period hours}} \times 100$

3. at generating end

4. $\dfrac{\text{Forced outage days}}{\text{Service days} + \text{Forced outage days}} \times 100$

5. Air dried basis

E.   Summary Comment

The Japanese and Australian data seem to suggest two ends of an economic spectrum. The Australian penalty for using inconsistent and out of design specification coals was high, probably higher than most utilities will experience under normal circumstances.

At the other end, the Japanese utilities have taken steps to reduce forced outages to as near zero as possible, probably closer than most utilities will normally experience.

In the final anlaysis the economic trade offs considering total system capacity availability, cost of coal (at various quality levels), maintenance costs, substitute fuel and capacity costs, plant replacement costs, etc. must be analysed for each operating situation. Only then can meaningful and specific conclusions about the cost impact of coal quality on the cost of electricity be made. Finally, judgements must be made trading off these costs versus qualitative factors such as diversity of supply, reliability, control of emissions for environmental reasons, balance of trade and currency availabilies in order to make final coal selection decisions.

Appendix A

PRINCIPAL INTERVIEWS CONDUCTED

United Kingdom

Central Electricity Generating Board
. John Wooley
. Arnold Shaw

National Coal Board
. I.K. MacGregor
. George Thurlow (CRE)
. John Ross Harrow

Babcock and Wilcox
. Dr. A. Sanyal

Ashton University
. Dr. Grines
. Dr. R.G. Temple

Foster Wheeler Power
. Dr. Beith

Denmark

      Elsam
         . H. Schulz

      Elkraft/IFV
         . Soren Nohr
         . Eric Odenburg
         . Mathias Christopherson
         . Erik Rosenberg

West Germany

      VDEW
         . Gustaf Heinman

      VGB
         . Dr. Joren Jacobs

Italy

      ENEL
         . Marco Gatti
         . P. Cerva

Holland

      IFRF
         . Peter Roberts

      PGEM
         . T.M. Hemels

Republic of South Africa

      ESCOM
         . Alec Hamm

United States

Combustion Engineering
. William H. Tuppeny, Jr.

Ebaso
. Eugene Chao

W.F. Berry and Associates
. Ralph Gray (retired U.S. Steel coal research expert)

Energy and Environmental Research Corporation
. Dr. Michael P. Heap
. Dr. Blair A. Folsom
. Dr. John A. Pohl

Applied Economic Research Company
. Dr. Marie R. Corio

U.S. Department of Commerce
. Dr. Joseph Jancik

Babcock and Wilcox
. Dr. A.W. Jackson

Electric Power Research Institute
. Arud Mehta

Canada

Combustion and Carbonisation Research Laboratory
Canadian Department of Energy, Mines and Resources
. George K. Lee

France

Electricité de France
. Lucien Edouard

Appendix B

SELECTED BIBLIOGRAPHY

1. MacKowsky:
   "The Mineral Constitution of Coal as the Causative Agent of the Fouling of Heating Surface with Particular Reference to Pulverised Fuel Firing with Liquid Slag Removal"; CEGB Marchwood Conference 1963, proceedings.

2. E. Raask:
   "Reactions of Coal Impurities During Combustion and Deposition of Ash Constituents on Cooled Surfaces"; CEGB Marchwood Conference 1963, proceedings.

3. D.C. Gunn:
   "The Effect of Coal Characteristics on Boiler Performance"; *Journal of the Institute of Fuel*, July 1952.

4. W.F. Simonson:
   "Boiler Design and Availability"; Boiler Availability Committee, London 1954.

5. Boiler Availability Committee:
   "Second Interim Report on External Deposits and Corrosion in Boiler Plants"; London 1953.

6. W.G. Marskell:
   "Developments in Cyclone Firing and Pulverised Fuel Firing of Low Volatile Coals"; Proceedings of the Second Conference of Pulverised Fuel, November 1957.

7. R.F. Davis:
    "Ten Years Development in Large PF-Fired Boilers"; Ibid.

8. K.E. Dadswell and F. Dransfield:
    "Some Design Requirements and Operating Experiences with Pulverised Fuel Milling Plant"; Ibid.

9. Lin Hao:
    "The Combination of Anthracites and Low Grade Bituminous Coals"; Proceedings of the 1983 International Conference on Coal Science, Pittsburgh, PA, United States.

10. A.H. Kirby:
    "The Influence of Ash Content on Boiler Plant Design and Performance"; Report of the British Coal Utilisation Research Association.

11. T.H. Geissler:
    "The Suitability of German and Foreign Fuels for Slag-Tap Firing"; *Energie*, 1960 (translated by British Coal Utilisation Research Association, Monthly Bulletin, March 1963).

12. T.H. Geissler:
    "The Suitability of Germans and Foreign Fuels for Slag-Tap Firing-Bituminous Coal — Part II"; Ibid, April 1963.

13. R.G. Bishop, K.R. Cliffe, T.H. Langford:
    "The Ash Deposition Characteristics of Seven British Coals"; BCURA Report DBA/32, December 1967.

14. D.R. Cooling:
    "Investigations of Wear in Coal Pulverisers"; BCURA Report No. 296, December 1965.

15. M.C. Christophersen:
    "Oil to Coal Conversion in Danish Power Station. A Case Study"; London Conference on Energy, 1982.

16. G. Tarjan:
    "Effect of Coal Quality on the Heating Efficiency of Furnaces" (copy available from British Museum), Hungarian Academy of Sciences, 1961.

17. G.K. Lee, W. Whaley:
"Modification of Combustion and Fly Ash Characteristics by Coal Blending"; *Journal of the Institute of Energy*, December 1983.

18. A. Sanyal, J. Williamson:
"Slagging in Boiler Furnaces: An Assessment Technique Based on Thermal Behaviour of Coal Minerals"; *Journal of the Institute of Energy*, September 1981.

19. A. Sanyal:
"The Role of Coal Macerals in Combustion"; *Journal of the Institute of Energy*, June 1983.

20. R.W. Brown:
"Relation Between Calorific Value and Ash Content for British Industrial Grade Coals"; *Journal of the Institute of Fuel*, November 1970.

21. W. Sgulakowsky:
"Influence of Silica on Some Properties of Coal Ash"; *Journal of the Institute of Fuel*, December 1967.

22. K. Schroeder:
"Possibilities to Reduce Power Generation Costs"; *Combustion*, August 1966 Vol. 38 No. 2.

23. A.C. Dunningham:
"Ash and Boiler Efficiency"; *Journal of the Institute of Fuel*, September 1952.

24. Council of Scientific and Industrial Research Organisations (CSIRO):
"Boiler-Fouling Problems with Leigh Creek Coal"; *Coal Research in CSIRO No. 28*, February 1966 (Australia).

25. H.E. Crossley:
"External Boiler Deposits"; *Journal of the Institute of Fuel*, September 1952.

26. S.G. Vecci, C.L. Wagoner, G.B. Olson:
"Fuel and Ash Characteristics and Its Effect on the Design of Industrial Boilers"; American Power Conference, April 1978.

27. J.E. Roughton:
"A Proposed On-Line Efficiency Method for Pulverised Coal-Fired Boilers"; *Journal of the Institute of Energy,* March 1980.

28. W.D. Halstead, E. Raask:
"The Behaviour of Sulphur and Chlorine Compounds in Pulverised Coal-Fired Boilers"; *Journal of the Institute of Fuel,* September 1969.

29. G. Wronski, R.G. Wilson:
"Modernisation of Control Systems to Maintain Plant Reliability and Economy in a Flexible Operating Regime"; *Proceedings Institute Mechanical Engineers* Vol. 197A, 1983.

30. Electric Power Research Institute:
"Coal Cleaning Test Facility Campaign Report Number 1"; April 1984.

31. G.W. Bouton, K.H. Haller, H.K. Smith:
"Ten Years Experience with Large Pulverised Coal-Fired Boilers for Utility Service": American Power Conference, April 1982 (available from Babcock and Wilcox, Barberton, Ohio).

32. T.C. Heil, O.W. Durant:
"Designing Boilers for Western Coal"; Joint Power Generation Conference, September 1978 (available from Babcock and Wilcox, Barberton, Ohio).

33. A.F. Armor et al:
"The Next Generation of Pulverised-Coal Power Plants"; American Power Conference, April 1981 (available from Babcock and Wilcox, Barberton, Ohio).

34. Bechtel National Inc.:
"Impact of Coal Cleaning on the Cost of New Coal-Fired Power Generation"; EPRI, March 1981.

35. Battelle-Columbus Laboratories:
"Examining Relationships Between Coal Characteristics and the Performance of TVA Power Plants — Final Report"; TVA/OP/EDT — 83/12, September 1982.

36. E. Chao:
    "Attainment of High Generating Unit Reliability and Availability"; Ebasco Services Inc., New York City.

37. C.D. Harrison:
    "Historical Deterioration of U.S. Coal Quality — Effects on the Power Industry"; Second Annual Seminar on Electric Utility Research and Development, October 1983.

38. Baur, Paul S.:
    "Coal Cleaning to Improve Boiler Performance and Reduce SO_2 Emmissions." *Power,* September 1981, Vol. 125, No. 9, 2-1—216.

40. Corio, Marie R. and Condran, Alice E.:
    "Which Coal at What Cost"; *Public Utilities Fortnightly,* March 1984.

41. Duzy, Albert F. and Pacer, Donald W.:
    "How Coal Quality Affects Boiler Design". *Coal Mining and Processing,* May 1982: 72-78.

42. Holt, Elmer C. Jr.:
    "Effect of Coal Quality on Maintenance Costs at Utility Plants". *Mining Congress Journal,* May 1982: 48-54.

43. Rittenhouse, R.C.:
    "Coal Supplies and the Economics of Treatment". *Power Engineering,* June 1982: 44-52.

44. Vaninetti, Gerald E. and Busch, C.F. 1982:
    "Mineral Analysis of Ash Data: A Utility Perspective". *Journal of Coal Quality,* Spring 1982: 22-31.

45. American Electric Power Service Corporation 1982:
    "Coal Quality and Steam Generation."

46. Blackmore, Gerald. 1981:
    "Coal Cleaning, Steam Raising and Environmental Control: An Economic Integration of Systems." Presented at Integrated Environmental Control for Coal-Fired Power Plants Symposium, Denver, Colorado.

47. Blackmore, Gerald. 1982:
    "Quality Coal for Electric Utility Use." Presented at Ninth Annual International Conference on Coal Gasification, Liquefaction and Conversion to Electricity, Pittsburgh, Pennsylvania.

48. Davidson, P.G. and Galluzzo, N.G. 1980:
    "Impact on Coal Constituents on Generation Plant Costs." Presented at Pacific Coast Electrical Association Engineering and Operating Conference, San Francisco, California.

49. Duzy, A.F. and Suydam, C.D. Jr. 1977:
    "An Economic Evaluation of Washed Coal for the Four Corners Generating Station." Presented at the Winter Annual Meeting of the American Society of Mechanical Engineers, Atlanta, Georgia.

50. Phillips, Peter J., and Cole, Randy, M. 1979:
    "Economic Penalties Attributable to Ash Content of Steam Coals." Presented at Coal Utilisation Symposium AIME Annual Meeting, New Orleans, Louisiana.

51. Vaninetti, G.E. and C.F. Busch. 1980:
    "The Effects of Beneficiation on the Performance of a Western Steam Coal." Presented at the Third International Coal Utilisation Conference, Houston, Texas.

52. Vaninetti, G.E. and C.F. Busch. 1981:
    "A Utility Perspective on the Significance of Mineral Analysis of Ash Data." Presented at the First Coal Testing Conference, Lexington, Kentucky.

53. Vaninetti, G.E. 1981:
    "Technical Considerations of Coal Supply Procurement." Presented at Applied Coal Geoscience and the Electric Utilities EPRI Workshop, Austin, Texas.

54. Burkhardt, Fred R. and Persinger, Marvin M.:
    "Economic Evaluation of Losses to Electric Power Utilities Caused by Ash Fouling: Final Technical Report No. 1979-1980." Prepared for U.S. Department of Energy.

55. Burkhardt, Fred R. and Persinger, Marvin M.:
"Economic Evaluation of Losses to Electric Power Utilities Caused by Ash Fouling: Executive Summary Report." Prepared for U.S. Department of Energy.

56. Buscheck, Timothy E.; Smith, Randall T.; and Burr, Myron W. III. 1981: "Economic Evaluation of Alternative Solutions to Decrease Boiler Ash Fouling." Final Report. Prepared for U.S. Department of Energy.

57. Isaacs, G.A.; Ress, R.A.; and Spaite, P.W.:
"Cost Benefits Associated with the Use of Physically Cleaned Coal." Prepared for U.S. Environmental Protection Agency, Office of Research and Development.

58. Buckler, Robert J.; Fletcher, Harold R.; Garbinski, Ernest H. and Adler, Ronald J. 1981:
"The Development of Optimum Economic and Technical Coal Specifications and Supplies." Presented at the 43rd Annual Meeting American Power Conference, Chicago, Illinois Detroit Edison.

Appendix C

GLOSSARY

The following glossary contains definitions or conversion factors for terms commonly used in steam coal literature.

Definition of Units

kilowatt hours (kWh) are 10^3 watt hours
megawatt hours (MWh) are 10^6 watt hours
gigawatt hours (GWh) are 10^9 watt hours
megajoules (MJ) are 10^6 joules
gigajoules (GJ) are 10^9 joules
1 therm = 100,000 Btu
1 metric ton of coal equivalent =   7,000 kcal/kg or
                             12,600 Btu/1b
                    = 7 million kcal or
                       27.8 million Btu

1 metric ton of oil equivalent = 10 million kcal
Mtce   = million (metric) tons of coal equivalent
Mtoe   = million (metric) tons of oil equivalent
Btu   = British Thermal Unit
kcal   = kilocalories (1000 calories)

Conversion Factors

1 megajoule = 947.81 Btu
238.85 kcal

1 gigajoule per metric tonne = 429.92 Btu/lb
238.85 kcal/kg

1 tonne = 2,204.6 lbs or 1.1 short ton

(Note: unless otherwise modified "ton" refers to the short, or net, ton of 2,000 lbs and "tonne" refers to the metric ton).

Heat Contents

Fuel oil   = 43.8 GJ/tonne, 6.77 barrels/tonne
Diesel fuel = 45.6 GJ/tonne, 7.43 barrels/tonne
Natural gas = 38.2 MJ/cubic metre

Energy Conversion

GJ/MWh	Efficiency
3.6	100%
7.2	50
10.3	35
12.0	30
14.4	25

OECD SALES AGENTS
DÉPOSITAIRES DES PUBLICATIONS DE L'OCDE

ARGENTINA - ARGENTINE
Carlos Hirsch S.R.L., Florida 165, 4° Piso (Galería Guemes)
1333 BUENOS AIRES, Tel. 33.1787.2391 y 30.7122

AUSTRALIA - AUSTRALIE
Australia and New Zealand Book Company Pty, Ltd.,
10 Aquatic Drive, Frenchs Forest, N.S.W. 2086
P.O. Box 459, BROOKVALE, N.S.W. 2100. Tel. (02) 452.44.11

AUSTRIA - AUTRICHE
OECD Publications and Information Center
4 Simrockstrasse 5300 Bonn (Germany). Tel. (0228) 21.60.45
Local Agent/Agent local :
Gerold and Co., Graben 31, WIEN 1. Tel. 52.22.35

BELGIUM - BELGIQUE
Jean De Lannoy, Service Publications OCDE
avenue du Roi 202, B-1060 BRUXELLES. Tel. 02/538.51.69

CANADA
Renouf Publishing Company Limited,
Central Distribution Centre,
61 Sparks Street (Mall),
P.O.B. 1008 - Station B,
OTTAWA, Ont. K1P 5R1.
Tel. (613)238.8985-6
Toll Free: 1-800.267.4164
Librairie Renouf Limitée
980 rue Notre-Dame,
Lachine, P.Q. H8S 2B9,
Tel. (514) 634-7088.

DENMARK - DANEMARK
Munksgaard Export and Subscription Service
35, Nørre Søgade
DK 1370 KØBENHAVN K. Tel. +45.1.12.85.70

FINLAND - FINLANDE
Akateeminen Kirjakauppa
Keskuskatu 1, 00100 HELSINKI 10. Tel. 65.11.22

FRANCE
Bureau des Publications de l'OCDE,
2 rue André-Pascal, 75775 PARIS CEDEX 16. Tel. (1) 524.81.67
Principal correspondant :
13602 AIX-EN-PROVENCE : Librairie de l'Université.
Tel. 26.18.08

GERMANY - ALLEMAGNE
OECD Publications and Information Center
4 Simrockstrasse 5300 BONN Tel. (0228) 21.60.45

GREECE - GRÈCE
Librairie Kauffmann, 28 rue du Stade,
ATHÈNES 132. Tel. 322.21.60

HONG-KONG
Government Information Services,
Publications/Sales Section, Baskerville House,
2nd Floor, 22 Ice House Street

ICELAND - ISLANDE
Snaebjörn Jönsson and Co., h.f.,
Hafnarstraeti 4 and 9, P.O.B. 1131, REYKJAVIK.
Tel. 13133/14281/11936

INDIA - INDE
Oxford Book and Stationery Co. :
NEW DELHI-1, Scindia House. Tel. 45896
CALCUTTA 700016, 17 Park Street. Tel. 240832

INDONESIA - INDONÉSIE
PDIN-LIPI, P.O. Box 3065/JKT., JAKARTA, Tel. 583467

IRELAND - IRLANDE
TDC Publishers - Library Suppliers
12 North Frederick Street, DUBLIN 1 Tel. 744835-749677

ITALY - ITALIE
Libreria Commissionaria Sansoni :
Via Lamarmora 45, 50121 FIRENZE. Tel. 579751/584468
Via Bartolini 29, 20155 MILANO. Tel. 365083
Sub-depositari :
Ugo Tassi
Via A. Farnese 28, 00192 ROMA. Tel. 310590
Editrice e Libreria Herder,
Piazza Montecitorio 120, 00186 ROMA. Tel. 6794628
Costantino Ercolano, Via Generale Orsini 46, 80132 NAPOLI. Tel. 405210
Libreria Hoepli, Via Hoepli 5, 20121 MILANO. Tel. 865446
Libreria Scientifica, Dott. Lucio de Biasio "Aeiou"
Via Meravigli 16, 20123 MILANO Tel. 807679
Libreria Zanichelli
Piazza Galvani 1/A, 40124 Bologna Tel. 237389
Libreria Lattes, Via Garibaldi 3, 10122 TORINO. Tel. 519274
La diffusione delle edizioni OCSE è inoltre assicurata dalle migliori librerie nelle
città più importanti.

JAPAN - JAPON
OECD Publications and Information Center,
Landic Akasaka Bldg., 2-3-4 Akasaka,
Minato-ku, TOKYO 107 Tel. 586.2016

KOREA - CORÉE
Pan Korea Book Corporation,
P.O. Box n° 101 Kwangwhamun, SÉOUL. Tel. 72.7369

LEBANON - LIBAN
Documenta Scientifica/Redico,
Edison Building, Bliss Street, P.O. Box 5641, BEIRUT.
Tel. 354429 - 344425

MALAYSIA - MALAISIE
University of Malaya Co-operative Bookshop Ltd.
P.O. Box 1127, Jalan Pantai Baru
KUALA LUMPUR. Tel. 577701/577072

THE NETHERLANDS - PAYS-BAS
Staatsuitgeverij, Verzendboekhandel,
Chr. Plantijnstraat 1 Postbus 20014
2500 EA S-GRAVENHAGE. Tel. nr. 070.789911
Voor bestellingen: Tel. 070.789208

NEW ZEALAND - NOUVELLE-ZÉLANDE
Publications Section,
Government Printing Office Bookshops:
AUCKLAND: Retail Bookshop: 25 Rutland Street,
Mail Orders: 85 Beach Road, Private Bag C.P.O.
HAMILTON: Retail: Ward Street,
Mail Orders, P.O. Box 857
WELLINGTON: Retail: Mulgrave Street (Head Office),
Cubacade World Trade Centre
Mail Orders: Private Bag
CHRISTCHURCH: Retail: 159 Hereford Street,
Mail Orders: Private Bag
DUNEDIN: Retail: Princes Street
Mail Order: P.O. Box 1104

NORWAY - NORVÈGE
J.G. TANUM A/S
P.O. Box 1177 Sentrum OSLO 1. Tel. (02) 80.12.60

PAKISTAN
Mirza Book Agency, 65 Shahrah Quaid-E-Azam, LAHORE 3.
Tel. 66839

PORTUGAL
Livraria Portugal, Rua do Carmo 70-74,
1117 LISBOA CODEX. Tel. 360582/3

SINGAPORE - SINGAPOUR
Information Publications Pte Ltd,
Pei-Fu Industrial Building,
24 New Industrial Road N° 02-06
SINGAPORE 1953, Tel. 2831786, 2831798

SPAIN - ESPAGNE
Mundi-Prensa Libros, S.A.
Castelló 37, Apartado 1223, MADRID-1. Tel. 275.46.55
Libreria Bosch, Ronda Universidad 11, BARCELONA 7.
Tel. 317.53.08, 317.53.58

SWEDEN - SUÈDE
AB CE Fritzes Kungl Hovbokhandel,
Box 16 356, S 103 27 STH, Regeringsgatan 12,
DS STOCKHOLM. Tel. 08/23.89.00
Subscription Agency/Abonnements:
Wennergren-Williams AB,
Box 13004, S104 25 STOCKHOLM.
Tel. 08/54.12.00

SWITZERLAND - SUISSE
OECD Publications and Information Center
4 Simrockstrasse 5300 BONN (Germany). Tel. (0228) 21.60.45
Local Agents/Agents locaux
Librairie Payot, 6 rue Grenus, 1211 GENÈVE 11. Tel. 022.31.89.50

TAIWAN - FORMOSE
Good Faith Worldwide Int'l Co., Ltd.
9th floor, No. 118, Sec. 2,
Chung Hsiao E. Road
TAIPEI. Tel. 391.7396/391.7397

THAILAND - THAILANDE
Suksit Siam Co., Ltd., 1715 Rama IV Rd,
Samyan, BANGKOK 5. Tel. 2511630

TURKEY - TURQUIE
Kültur Yayinlari Is-Türk Ltd. Sti.
Atatürk Bulvari No : 191/Kat. 21
Kavaklidere/ANKARA. Tel. 17 02 66
Dolmabahce Cad. No : 29
BESIKTAS/ISTANBUL. Tel. 60 71 88

UNITED KINGDOM - ROYAUME-UNI
H.M. Stationery Office,
P.O.B. 276, LONDON SW8 5DT.
(postal orders only)
Telephone orders: (01) 622.3316, or
49 High Holborn, LONDON WC1V 6 HB (personal callers)
Branches at: EDINBURGH, BIRMINGHAM, BRISTOL,
MANCHESTER, BELFAST.

UNITED STATES OF AMERICA - ÉTATS-UNIS
OECD Publications and Information Center, Suite 1207,
1750 Pennsylvania Ave., N.W. WASHINGTON, D.C.20006 - 4582
Tel. (202) 724.1857

VENEZUELA
Libreria del Este, Avda. F. Miranda 52, Edificio Galipan,
CARACAS 106. Tel. 32.23.01/33.26.04/31.58.38

YUGOSLAVIA - YOUGOSLAVIE
Jugoslovenska Knjiga, Knez Mihajlova 2, P.O.B. 36, BEOGRAD.
Tel. 621.992

Les commandes provenant de pays où l'OCDE n'a pas encore désigné de dépositaire peuvent être adressées à :
OCDE, Bureau des Publications, 2, rue André-Pascal, 75775 PARIS CEDEX 16.

Orders and inquiries from countries where sales agents have not yet been appointed may be sent to:
OECD, Publications Office, 2, rue André-Pascal, 75775 PARIS CEDEX 16.

68236-12-1984

OECD PUBLICATIONS, 2, rue André-Pascal, 75775 PARIS CEDEX 16 - No. 43181 1985
PRINTED IN FRANCE
(61 85 02 1) ISBN 92-64-12684-8